EMMANUEL JOSEPH

The Science of Love: Exploring the Neurochemistry Behind Romantic Relationships

Copyright © 2023 by Emmanuel Joseph

All rights reserved. No part of this publication may be reproduced, stored or transmitted in any form or by any means, electronic, mechanical, photocopying, recording, scanning, or otherwise without written permission from the publisher. It is illegal to copy this book, post it to a website, or distribute it by any other means without permission.

First edition

*This book was professionally typeset on Reedsy.
Find out more at reedsy.com*

Contents

1	Chapter 1: Introduction to Love	1
2	Chapter 2: Evolutionary Basis of Love	3
3	Chapter 3: Brain in Love	5
4	Chapter 4: Chemistry of Attraction	7
5	Chapter 5: Love and Emotions	9
6	Chapter 6: The Role of Oxytocin in Relationships	11
7	Chapter 7: Love, Sex, and the Brain	13
8	Chapter 8: Attachment Styles and Relationships	15
9	Chapter 9: Love and Long-Term Relationships	17
10	Chapter 10: Love, Breakups, and Heartache	19
11	Chapter 11: Love and Mental Health	21
12	Chapter 12: The Future of Love: Neuroscientific Perspectives	23

1

Chapter 1: Introduction to Love

Love, an enigmatic and multifaceted emotion, has fascinated humanity throughout history. It's a feeling that transcends cultural boundaries, time periods, and societal norms, existing in various forms—romantic, familial, platonic—yet retaining a profound impact on individuals and societies alike. This chapter serves as a gateway to understanding the intricate neural mechanisms that underpin the phenomenon of love.

Defining Love: Historical and Cultural Perspectives

The definition of love is as diverse as the emotions it encapsulates. From ancient philosophical discourses to contemporary scientific inquiries, love has been contemplated and articulated in numerous ways across different cultures and eras. Early philosophical texts from ancient Greece to Eastern philosophies offer insights into the conceptualization of love as eros (romantic love), agape (unconditional love), or philia (platonic love).

Throughout history, literature, art, and religious scriptures have depicted love as a driving force that shapes human experiences, relationships, and societal structures. From Shakespearean sonnets to modern-day pop culture, the portrayal of love encompasses a spectrum of emotions, complexities, and expressions.

Different Types of Love

Love manifests in various forms and contexts, each with its unique characteristics and significance. Romantic love, often associated with passion, desire, and intense emotional connections, is one of the most explored types of love. Familial love, exemplified in the bonds between parents and children or among siblings, represents a deep-rooted connection based on shared experiences and genetic ties. Platonic love, characterized by camaraderie, trust, and non-sexual affection, enriches friendships and social connections.

Understanding the diversity of love allows us to appreciate its complexities and variations across relationships and cultures. From the fervor of newfound romance to the enduring bonds of lifelong companionship, love permeates our lives in multifaceted ways.

Overview of the Neurochemistry of Love

Advancements in neuroscience have unveiled the intricate neurobiological mechanisms underlying love, unraveling its biological underpinnings. The brain, the epicenter of our thoughts, emotions, and behaviors, orchestrates the symphony of love through a complex interplay of neurotransmitters, hormones, and brain regions.

This chapter sets the stage for a deeper exploration into the fascinating world of love from a neuroscientific standpoint. It invites readers on a journey to unravel the mysteries of the brain in love, offering a glimpse into the evolutionary origins, neurochemical substrates, and psychological dimensions that define our experiences and understanding of this profound emotion.

2

Chapter 2: Evolutionary Basis of Love

Love, often considered a cornerstone of human existence, finds its roots embedded within the evolutionary history of our species. Understanding the evolutionary underpinnings of love provides crucial insights into its purpose, functions, and adaptive significance in shaping human behavior and relationships.

Evolutionary Theories Behind the Need for Romantic Relationships

Evolutionary psychology posits that love, particularly romantic love, is not merely a byproduct of human culture but has evolved as an adaptive mechanism. Theories such as parental investment theory and sexual selection shed light on the evolutionary imperatives driving human mating behaviors and the formation of romantic relationships.

Parental investment theory suggests that the differing levels of investment in reproduction between males and females drive certain mating behaviors. Males, with lower investment due to sperm production, might seek multiple partners, while females, with higher investment in gestation and child-rearing, might prioritize selecting a suitable mate. This theory helps understand mate preferences and relationship dynamics.

Sexual selection, as proposed by Charles Darwin, elucidates how traits contributing to mating success are selected and passed down through generations. The display of certain behaviors or characteristics, such as physical attractiveness, resources, or social status, might confer advantages in attracting potential partners, contributing to the perpetuation of these traits.

Role of Love in Human Survival and Reproduction

Love, intertwined with the processes of mate selection, pair bonding, and parental care, plays a pivotal role in the survival and reproductive success of our species. It fosters attachment between partners, promoting cooperation, caregiving, and support, which enhance the chances of offspring survival.

The evolutionary significance of love extends beyond reproduction, influencing social cohesion and group dynamics. Strong social bonds forged through love and cooperation contribute to the success of communities and societies, enhancing collective survival and thriving.

Comparative Perspectives on Love in Different Species

While humans exhibit complex forms of love, similar social bonding mechanisms exist across various species. Observations in animals, such as monogamous pair bonding in certain bird species or social bonding in primates, highlight the evolutionary roots of social attachment and affiliation.

Studying love-like behaviors in animals provides valuable insights into the evolutionary origins and neural mechanisms of social bonding, shedding light on the shared biological foundations of love across diverse species.

This chapter delves into the evolutionary framework that underlies the existence of love in humans, exploring how natural selection has shaped our behaviors, emotions, and social connections, ultimately contributing to the survival and flourishing of our species.

3

Chapter 3: Brain in Love

The intricate dance of love is orchestrated within the complex terrain of the human brain. This chapter delves into the neural substrates, neurotransmitters, and brain regions that form the foundation of love's emotional and cognitive dimensions.

Brain Regions and Neurotransmitters Involved in Love

Neuroscience has unveiled a network of brain regions that play a pivotal role in the experience of love. The limbic system, particularly the amygdala, hippocampus, and hypothalamus, is implicated in processing emotions, forming memories, and regulating physiological responses associated with love, attachment, and social bonding.

Neurotransmitters serve as chemical messengers orchestrating the neurochemical symphony of love. Dopamine, often referred to as the "pleasure molecule," plays a central role in the brain's reward system, contributing to feelings of pleasure, desire, and motivation. Serotonin regulates mood and emotional processing, while oxytocin and vasopressin, known as bonding hormones, foster attachment, trust, and social bonding.

Neurobiology of Attraction and Attachment

The stages of attraction, attachment, and long-term bonding involve distinct neural pathways and neurochemical processes. During the initial phase of attraction, heightened activity in reward-related brain regions, such as the ventral tegmental area (VTA) and nucleus accumbens, is observed. This surge in dopamine levels contributes to the euphoria and obsession often experienced in new romantic relationships.

Attachment, marked by feelings of security and emotional closeness, is facilitated by the release of oxytocin and vasopressin, promoting pair bonding and social attachment. These hormones modulate the activity of brain regions associated with trust, empathy, and social bonding, fostering enduring connections between partners.

Effects of Love on Brain Structure and Function

Neuroplasticity—the brain's ability to rewire and adapt—is evident in the context of love. Long-term romantic relationships and experiences of love can induce structural changes in the brain. Studies have shown alterations in brain areas associated with empathy, emotional regulation, and social cognition, reflecting the profound impact of love on the brain's architecture and functionality.

Understanding the neural mechanisms underlying love provides a fascinating glimpse into the intricate interplay between emotions, cognition, and biology. This exploration into the brain in love elucidates how our neural circuitry shapes and responds to the profound emotions and connections experienced in romantic relationships.

4

Chapter 4: Chemistry of Attraction

The intoxicating allure of attraction, a magnetic force drawing individuals together, is governed by a complex interplay of hormones, neurotransmitters, and behavioral cues. This chapter unravels the neurochemical underpinnings of attraction, exploring its phases and the biological factors that influence this captivating phenomenon.

Role of Hormones in Attraction

Attraction, the initial stage of romantic love, involves a surge of hormones that contribute to the feelings of exhilaration and infatuation experienced in the presence of a potential partner. Dopamine, known for its role in pleasure and reward, floods the brain during this phase, generating feelings of excitement and anticipation. Increased dopamine levels drive motivation and reinforce the pursuit of a romantic interest.

Oxytocin, often termed the "love hormone," and serotonin also play pivotal roles in attraction. Oxytocin fosters feelings of trust and bonding, while serotonin regulates mood and emotional balance, influencing the intensity and stability of romantic attraction.

Phases of Attraction: Lust, Attraction, Attachment

Attraction unfolds in distinct phases—lust, attraction, and attachment—each characterized by unique hormonal and neurobiological signatures. Lust, driven primarily by the hormone testosterone, encompasses the initial physical desire and sexual attraction toward a potential partner.

The attraction phase, marked by the euphoric feelings of infatuation and obsession, involves the interplay of dopamine, norepinephrine, and serotonin. This phase fuels the intense focus on a romantic interest, often leading to heightened emotional arousal and preoccupation with the object of desire.

Attachment, the stage characterized by feelings of security and emotional closeness, is primarily mediated by oxytocin and vasopressin. These bonding hormones facilitate the formation of long-term connections and deepen emotional intimacy between partners.

Genetic and Biological Factors Influencing Attraction

Biological and genetic factors significantly influence individual differences in attraction and mate preferences. Genetic compatibility, olfactory cues, pheromones, and immune system compatibility contribute to the subconscious evaluation of potential partners.

The human leukocyte antigen (HLA) genes, for instance, influence the perception of body odor, impacting attraction and mate selection. Evolutionarily, diverse HLA genes in partners might confer advantages in offspring immunity.

This exploration into the chemistry of attraction unveils the intricate interplay between hormones, genetics, and biological cues that shape the initial stages of romantic love. Understanding the neurochemical mechanisms of attraction provides insights into the complexities of mate selection and the compelling forces that drive individuals toward romantic connections.

5

Chapter 5: Love and Emotions

Love, an intricate tapestry of emotions, intertwines joy, anxiety, contentment, and vulnerability. This chapter delves into the emotional landscape of love, exploring the interplay between love and various emotional states, the impact of emotions on relationships, and strategies for emotional regulation within romantic bonds.

Emotional Spectrum of Love

Love encompasses a wide spectrum of emotions that fluctuate throughout the course of a relationship. Joy and happiness often accompany the initial stages of romantic love, characterized by euphoria, excitement, and intense passion. However, love's emotional terrain also navigates through periods of uncertainty, anxiety, jealousy, and vulnerability.

Anxiety in love may stem from fear of rejection, concerns about the stability of the relationship, or insecurities about one's own worthiness as a partner. Jealousy, another complex emotion, emerges from perceived threats to the relationship or feelings of possessiveness.

Impact of Emotions on the Neurochemistry of Love

Emotions wield a profound influence on the neurochemical dynamics of love. Positive emotions, such as happiness and contentment, contribute to the release of neurotransmitters like dopamine and serotonin, reinforcing feelings of connection and satisfaction within the relationship.

Conversely, negative emotions like anxiety or jealousy trigger stress responses in the brain, affecting the balance of neurochemicals and potentially straining the relationship. Persistent negative emotions may lead to increased cortisol levels, impacting emotional well-being and relationship harmony.

Emotional Regulation in Relationships

Effectively managing emotions is crucial for fostering healthy and resilient relationships. Strategies for emotional regulation, such as open communication, empathy, active listening, and conflict resolution skills, play pivotal roles in navigating the emotional landscape of love.

Mindfulness practices, emotional validation, and cultivating self-awareness aid in recognizing and addressing emotional triggers. Couples who develop emotional intelligence and coping mechanisms are better equipped to navigate challenges and nurture intimacy within their relationships.

Understanding the intricate interplay between love and emotions offers valuable insights into the complexities of romantic relationships. Embracing the diverse emotional experiences within love and adopting healthy emotional regulation strategies empower individuals to foster deeper connections and emotional intimacy within their relationships.

6

Chapter 6: The Role of Oxytocin in Relationships

Central to the fabric of human relationships, oxytocin—the "bonding hormone"—weaves threads of trust, intimacy, and social connection, profoundly influencing the dynamics and depth of interpersonal bonds. This chapter explores the pivotal role of oxytocin in fostering attachment, trust, and the intricacies of relationship dynamics.

Oxytocin: The Neurochemical Foundation of Bonding

Oxytocin, produced in the hypothalamus and released by the pituitary gland, serves as a key regulator of social behavior and emotional bonding. Often associated with maternal bonding during childbirth and breastfeeding, oxytocin extends its influence beyond parent-child relationships to encompass various forms of social interactions, including romantic relationships, friendships, and social affiliations.

Effects of Oxytocin on Bonding and Trust

The release of oxytocin plays a crucial role in nurturing feelings of trust, empathy, and attachment within relationships. Studies have demonstrated that elevated levels of oxytocin contribute to increased trust between partners, fostering emotional closeness and enhancing relationship satisfaction.

Oxytocin not only promotes bonding between romantic partners but also influences social behaviors, such as altruism and cooperation, by enhancing prosocial tendencies. Its role in modulating social interactions and promoting positive social behaviors underscores its significance in fostering harmonious relationships.

Oxytocin's Influence on Relationship Dynamics

The levels of oxytocin in individuals can impact relationship dynamics and responsiveness to social cues. Oxytocin's effects on enhancing empathy and reducing stress may contribute to better conflict resolution, emotional regulation, and mutual understanding within relationships.

Furthermore, individual differences in oxytocin receptor gene variations can influence oxytocin's effects on social behavior and attachment. Variations in receptor genes may affect sensitivity to oxytocin's effects, potentially shaping an individual's social tendencies and relationship patterns.

Understanding the intricate role of oxytocin in fostering trust, attachment, and social bonding sheds light on the neurobiological foundations of intimate relationships. Exploring the mechanisms through which oxytocin influences relationship dynamics provides valuable insights into enhancing closeness, trust, and the overall quality of interpersonal connections.

7

Chapter 7: Love, Sex, and the Brain

The intimate intertwining of love and sex represents a fascinating interplay of neurochemistry, emotions, and physiological responses within the intricate landscape of the human brain. This chapter delves into the neuroscientific underpinnings of sexual attraction, arousal, and the profound impact of sex on romantic relationships.

Neurochemical Basis of Sexual Attraction and Intimacy

Sexual attraction, a fundamental aspect of romantic relationships, involves a complex interplay of neurochemicals and brain regions. Dopamine, oxytocin, and testosterone contribute to the experience of sexual desire and attraction, stimulating feelings of pleasure, intimacy, and motivation.

Brain imaging studies reveal the activation of reward-related brain regions, such as the nucleus accumbens and prefrontal cortex, during sexual arousal. These regions play crucial roles in processing pleasure and anticipation, amplifying the emotional and physiological responses to sexual stimuli.

Brain Mechanisms During Sexual Arousal and Orgasm

Sexual arousal triggers a cascade of neural events within the brain, leading to increased blood flow to genital regions and heightened arousal states. Brain imaging studies have identified the involvement of various brain regions, including the hypothalamus, amygdala, and insula, during different phases of sexual response.

The climax of sexual arousal, the orgasm, is accompanied by a surge of neural activity and the release of oxytocin and endorphins. This release contributes to feelings of intense pleasure, emotional bonding, and the potential reinforcement of pair bonding within intimate relationships.

Interplay Between Love, Sex, and Relationship Satisfaction

The connection between love, sex, and relationship satisfaction underscores the intricate nature of romantic bonds. The frequency and quality of sexual intimacy within relationships are linked to overall relationship satisfaction, emotional closeness, and the maintenance of long-term connections.

Research suggests that the neurochemical responses associated with sexual activity contribute to enhanced emotional bonding and relationship stability. Moreover, the emotional intimacy fostered through sexual encounters can reinforce the feelings of attachment and trust between partners.

Understanding the neurobiological mechanisms of sexual attraction and intimacy provides valuable insights into the complex interplay between love and physical intimacy within romantic relationships. Exploring the brain's responses to sexual stimuli sheds light on the profound effects of sexual experiences on emotional bonding and relationship dynamics.

8

Chapter 8: Attachment Styles and Relationships

Human relationships are shaped by attachment patterns formed early in life and later influenced by experiences and interactions within intimate bonds. This chapter explores attachment theory, the various attachment styles, and their profound impact on relationship dynamics and the neurobiology of love.

Overview of Attachment Theory

Attachment theory, developed by John Bowlby and expanded by Mary Ainsworth, posits that early experiences with caregivers shape individuals' internal working models of relationships. These models influence how individuals perceive, approach, and engage in relationships throughout their lives.

The attachment behavioral system, essential for survival, seeks proximity to caregivers for protection, comfort, and security. This innate system lays the groundwork for emotional regulation, exploration, and the development of trust within relationships.

Different Attachment Styles and Their Neurobiological Underpinnings

Attachment styles—secure, anxious, avoidant, and disorganized—reflect individuals' expectations and behaviors in close relationships. Secure attachment, characterized by trust, comfort with intimacy, and effective communication, is associated with positive relationship outcomes.

Anxious attachment stems from a fear of abandonment and often manifests as excessive need for reassurance or preoccupation with the relationship. Avoidant attachment involves a discomfort with intimacy, independence, and a tendency to distance oneself emotionally.

Neuroscience studies reveal distinct neural correlates associated with different attachment styles. Brain imaging studies demonstrate variations in brain activity and responsiveness to social cues among individuals with different attachment patterns, highlighting the neural underpinnings of attachment styles.

Influence of Attachment Styles on Relationship Dynamics

Attachment styles significantly impact relationship dynamics, communication, and conflict resolution strategies. Partners with compatible attachment styles often experience greater relationship satisfaction and emotional intimacy. In contrast, mismatched attachment patterns may lead to misunderstandings, conflicts, and relationship distress.

Understanding one's attachment style and its influence on relationship behaviors enables individuals to recognize and address underlying patterns, fostering healthier relationship dynamics and enhancing emotional connections.

Exploring the neurobiological basis of attachment styles offers valuable insights into the mechanisms through which early experiences shape relationship patterns and the profound impact of attachment styles on the dynamics and quality of intimate relationships.

9

Chapter 9: Love and Long-Term Relationships

The evolution of love within long-term relationships traverses a terrain marked by companionship, commitment, and the interplay of neurochemical processes that sustain enduring bonds. This chapter delves into the neurochemistry of long-term commitment, factors contributing to lasting love, and strategies for maintaining intimacy within enduring relationships.

Neurochemistry of Long-Term Commitment and Companionship

Long-term relationships undergo transitions from passionate love to companionate love, characterized by deep emotional intimacy, trust, and companionship. The neurobiology of enduring love involves the modulation of neurotransmitters and hormones that facilitate bonding and emotional closeness.

Oxytocin and vasopressin continue to play pivotal roles in sustaining attachment and trust within long-term relationships. The release of these bonding hormones fosters emotional connections, promotes mutual support, and contributes to relationship stability over time.

Factors Contributing to Lasting Love in Relationships

Various factors contribute to the resilience and longevity of relationships. Effective communication, shared values, mutual respect, and the ability to adapt to changing circumstances are essential components that strengthen the foundation of enduring love.

Couples who engage in shared activities, maintain intimacy, and prioritize emotional connection often report higher relationship satisfaction. Additionally, the ability to navigate challenges, overcome conflicts, and grow together fosters resilience within relationships.

Challenges and Strategies for Maintaining Love in Long-Term Relationships

Long-term relationships encounter challenges, including routine, life stressors, and potential decreases in passion. However, intentional efforts to reignite passion, nurture emotional intimacy, and prioritize the relationship amidst life's demands contribute to relationship longevity.

Practices such as expressing gratitude, fostering novelty and excitement, and engaging in mutual support and appreciation help sustain the emotional connection between partners. Continual efforts to strengthen emotional bonds and rekindle the spark of romance are vital for nurturing enduring love.

Understanding the neurochemical mechanisms underpinning long-term commitment and exploring strategies for fostering intimacy in enduring relationships offers valuable insights into the dynamics of lasting love. This exploration delves into the complexities of maintaining emotional closeness and fostering mutual support within the context of lifelong partnerships.

10

Chapter 10: Love, Breakups, and Heartache

The dissolution of romantic relationships—marked by heartache, emotional turmoil, and neurobiological shifts—holds profound implications for individuals' well-being and emotional health. This chapter examines the neurochemical changes during breakups, the effects of heartbreak on the brain and body, and strategies for coping and healing after relationship loss.

Neurochemical Changes During Relationship Dissolution

Breakups trigger significant neurochemical alterations within the brain, akin to withdrawal from addictive substances. The brain regions associated with reward, including the nucleus accumbens and prefrontal cortex, exhibit decreased activity, leading to feelings of withdrawal, sadness, and emotional pain.

Dopamine, the neurotransmitter associated with pleasure and reward, experiences fluctuations during breakups. The absence of constant rewards from the relationship may lead to dopamine dysregulation, contributing to feelings of depression and withdrawal symptoms.

Effects of Heartbreak on the Brain and Body

Heartbreak exerts tangible effects on both the brain and the body. Studies using neuroimaging techniques have demonstrated changes in brain activity, particularly in regions associated with emotional regulation and pain perception, akin to physical pain experiences.

The body's stress response is activated during heartbreak, leading to increased cortisol levels and physiological stress. Prolonged stress may compromise the immune system, impacting overall health and well-being.

Coping Mechanisms and Recovery After a Breakup

Coping with the aftermath of a breakup involves adopting strategies to navigate emotional distress and facilitate healing. Engaging in self-care practices, seeking social support, and allowing oneself to grieve the loss are essential steps in the healing process.

Practicing mindfulness, engaging in hobbies, and focusing on personal growth contribute to emotional resilience and facilitate the transition towards acceptance and healing. Redirecting attention to self-discovery and rediscovering personal identity fosters emotional recovery and growth post-breakup.

Understanding the neurobiological ramifications of breakups, acknowledging the emotional upheaval experienced during heartache, and employing effective coping strategies facilitate the healing process. This exploration offers insights into the emotional complexities surrounding relationship loss and provides guidance for individuals seeking to navigate the challenging terrain of heartbreak.

11

Chapter 11: Love and Mental Health

The intricate interplay between love and mental health is a multifaceted terrain that influences individual well-being within the context of relationships. This chapter investigates the impact of love and relationships on mental health, the neurochemical basis of love-related mental health benefits, and approaches to addressing mental health challenges within intimate bonds.

Impact of Love and Relationships on Mental Well-Being

Healthy and supportive relationships significantly contribute to positive mental health outcomes. Strong social connections, emotional support, and feelings of belongingness within relationships foster resilience against stress, anxiety, and depression.

Conversely, unhealthy relationships characterized by conflict, emotional neglect, or toxicity can exert adverse effects on mental health. Relationship distress, loneliness, and unresolved conflicts may contribute to heightened levels of stress and emotional instability.

Neurochemical Basis of Love-Related Mental Health Benefits

The neurochemistry of love and relationships plays a crucial role in mental well-being. Positive social interactions and emotional closeness within relationships stimulate the release of oxytocin, dopamine, and serotonin, contributing to feelings of happiness, contentment, and stress reduction.

Oxytocin's role in fostering trust and bonding within relationships not only enhances emotional connection but also contributes to overall mental resilience. Dopamine's involvement in reward processing and pleasure reinforces positive feelings associated with healthy relationships, promoting emotional stability and mental well-being.

Addressing Mental Health Challenges Within Relationships

Recognizing and addressing mental health challenges within relationships is paramount for maintaining overall well-being. Effective communication, empathy, and support are essential in creating a safe space for discussing mental health concerns within intimate bonds.

Seeking professional help, such as therapy or counseling, can provide valuable tools and strategies for addressing mental health issues within relationships. Couples therapy or individual counseling offers opportunities to navigate challenges, improve communication, and foster emotional well-being.

Understanding the reciprocal relationship between love, relationships, and mental health sheds light on the profound influence of intimate bonds on individual well-being. This exploration illuminates the ways in which healthy relationships can contribute to mental resilience while emphasizing the importance of addressing mental health challenges within the context of love and intimacy.

12

Chapter 12: The Future of Love: Neuroscientific Perspectives

The evolving landscape of neuroscience presents an intriguing pathway for unraveling the mysteries of love, offering promising avenues for future research and applications. This chapter explores the advancing frontiers of neuroscientific inquiry into love, ethical considerations, and potential future developments shaping our understanding and application of the science of love.

Advances in Neuroscience and Love Research

Continual advancements in neuroimaging techniques, such as functional MRI (fMRI) and neurochemical assays, provide unprecedented insights into the neurobiological substrates of love. Cutting-edge research methodologies allow for a more comprehensive exploration of the brain mechanisms underlying emotions, social bonding, and relationship dynamics.

Emerging interdisciplinary studies combining neuroscience, genetics, psychology, and social sciences offer holistic perspectives on the complexities of love. Deepening our understanding of the interplay between genetics, environment, and neurobiology broadens the scope of love research, paving the way for comprehensive insights into human relationships.

Ethical Considerations in Studying the Neurochemistry of Love

The exploration of love through neuroscience raises ethical considerations regarding privacy, consent, and potential applications. Ethical frameworks guiding research on intimate emotions like love emphasize the importance of informed consent, confidentiality, and respect for individuals' autonomy in participating in studies.

Navigating ethical boundaries in utilizing neuroscientific knowledge about love requires careful consideration of potential implications. Ethical guidelines aim to balance the pursuit of knowledge with the ethical responsibilities to protect individuals' rights and prevent exploitation or misuse of research findings.

Potential Future Developments in Love Research and Applications

The integration of neuroscience with technology, such as wearable devices or neurofeedback interventions, holds promise for enhancing relationship dynamics and emotional well-being. Applications leveraging neuroscientific insights may offer innovative approaches for relationship counseling, improving communication, and fostering emotional connection.

Future research endeavors might delve deeper into personalized approaches to understanding individual differences in love-related neurobiology. Tailored interventions based on neuroscientific assessments could offer personalized strategies for enhancing relationship satisfaction and emotional health.

As we navigate the frontiers of neuroscience in unraveling the science of love, ethical considerations and the responsible application of knowledge remain paramount. The ongoing quest to decode the neural intricacies of love holds vast potential for shaping the future landscape of relationships, emotional well-being, and societal dynamics.

Book Description

"Unraveling Love: Exploring the Neurochemistry Behind Romantic Relationships"

"Unraveling Love" takes readers on a captivating journey through the intricate landscape of human emotions, relationships, and the fascinating world of neurochemistry. Delving into the depths of the human brain, this groundbreaking book explores the neural mechanisms that underpin the

CHAPTER 12: THE FUTURE OF LOVE: NEUROSCIENTIFIC PERSPECTIVES

enigmatic phenomenon of love, offering a comprehensive understanding of love's neurobiological foundations.

From the evolutionary origins of love to the neurochemical dynamics of attraction, attachment, and long-term commitment, each chapter navigates the complex terrain of romantic relationships. Readers discover the interplay of neurotransmitters like dopamine, oxytocin, and serotonin, unraveling their roles in shaping emotions, desires, and the intricate dance of intimacy.

This engaging exploration delves into the impact of emotions on relationships, the neurobiological basis of attachment styles, and the profound interconnections between love, sex, and the brain. Through in-depth discussions, readers gain insights into the neuroscience behind heartache, mental health implications within relationships, and strategies for fostering enduring love and emotional resilience.

"Unraveling Love" culminates in a visionary chapter, envisioning the future of love research through neuroscientific perspectives. Ethical considerations, potential advancements, and applications at the intersection of neuroscience and relationships offer glimpses into the promising future landscapes of emotional well-being and interpersonal connections.

Drawing on cutting-edge neuroscience and interdisciplinary research, "Unraveling Love" provides a captivating narrative that bridges the gap between science and emotion, offering a profound understanding of the intricate neurochemistry behind the most profound of human emotions: love.

Keywords

Love; Neurochemistry; Romantic Relationships; Neuroscience; Attachment Theory; Brain Chemistry; Oxytocin; Dopamine; Serotonin; Emotional Dynamics; Relationship Science; Evolutionary Psychology; Mental Health; Breakups; Emotional Resilience; Ethical Considerations; Future of Relationships; Interdisciplinary Research; Personalized Interventions; Emotional Well-being; Neuroscientific Perspectives; Communication; Intimacy.

Milton Keynes UK
Ingram Content Group UK Ltd.
UKHW050225130724
445574UK00013B/679